Strategic Studies Institute
and
U.S. Army War College Press

USING TARGET AUDIENCE ANALYSIS TO AID STRATEGIC LEVEL DECISIONMAKING

Steve Tatham

August 2015

The United States Army War College

The United States Army War College educates and develops leaders for service at the strategic level while advancing knowledge in the global application of Landpower.

The purpose of the United States Army War College is to produce graduates who are skilled critical thinkers and complex problem solvers. Concurrently, it is our duty to the U.S. Army to also act as a "think factory" for commanders and civilian leaders at the strategic level worldwide and routinely engage in discourse and debate concerning the role of ground forces in achieving national security objectives.

The Strategic Studies Institute publishes national security and strategic research and analysis to influence policy debate and bridge the gap between military and academia.

The Center for Strategic Leadership and Development contributes to the education of world class senior leaders, develops expert knowledge, and provides solutions to strategic Army issues affecting the national security community.

The Peacekeeping and Stability Operations Institute provides subject matter expertise, technical review, and writing expertise to agencies that develop stability operations concepts and doctrines.

The Senior Leader Development and Resiliency program supports the United States Army War College's lines of effort to educate strategic leaders and provide well-being education and support by developing self-awareness through leader feedback and leader resiliency.

The School of Strategic Landpower develops strategic leaders by providing a strong foundation of wisdom grounded in mastery of the profession of arms, and by serving as a crucible for educating future leaders in the analysis, evaluation, and refinement of professional expertise in war, strategy, operations, national security, resource management, and responsible command.

The U.S. Army Heritage and Education Center acquires, conserves, and exhibits historical materials for use to support the U.S. Army, educate an international audience, and honor Soldiers—past and present.

STRATEGIC STUDIES INSTITUTE

The Strategic Studies Institute (SSI) is part of the U.S. Army War College and is the strategic-level study agent for issues related to national security and military strategy with emphasis on geostrategic analysis.

The mission of SSI is to use independent analysis to conduct strategic studies that develop policy recommendations on:

- Strategy, planning, and policy for joint and combined employment of military forces;
- Regional strategic appraisals;
- The nature of land warfare;
- Matters affecting the Army's future;
- The concepts, philosophy, and theory of strategy; and,
- Other issues of importance to the leadership of the Army.

Studies produced by civilian and military analysts concern topics having strategic implications for the Army, the Department of Defense, and the larger national security community.

In addition to its studies, SSI publishes special reports on topics of special or immediate interest. These include edited proceedings of conferences and topically oriented roundtables, expanded trip reports, and quick-reaction responses to senior Army leaders.

The Institute provides a valuable analytical capability within the Army to address strategic and other issues in support of Army participation in national security policy formulation.

*****

Comments pertaining to this report are invited and should be forwarded to: Director, Strategic Studies Institute and U.S. Army War College Press, U.S. Army War College, 47 Ashburn Drive, Carlisle, PA 17013-5010.

*****

This manuscript was funded by the U.S. Army War College External Research Associates Program. Information on this program is available on our website, *www.StrategicStudiesInstitute.army.mil,* at the Opportunities tab.

*****

*****

The Strategic Studies Institute and U.S. Army War College Press publishes a monthly email newsletter to update the national security community on the research of our analysts, recent and forthcoming publications, and upcoming conferences sponsored by the Institute. Each newsletter also provides a strategic commentary by one of our research analysts. If you are interested in receiving this newsletter, please subscribe on the SSI website at *www.StrategicStudiesInstitute.army.mil/newsletter.*

ISBN 1-58487-699-9

FOREWORD

Albert Einstein famously stated that: "Any fool can know; the point is to understand." Over the past 20 years, the United States has known that there exist people with a profound hatred of all that it and the West are, and all that it stands for. During that time the American people and our allies abroad have known war in Iraq, in Afghanistan, and we know that today in Syria, Iraq, in Nigeria and North Africa those enemies plan and plot more violence and more hatred against us. Yet, do we really understand? If there is one observation that has been repeated by military commanders and policymakers alike from almost every nation in our various coalitions, it is the idea that we have not understood our adversary properly.

Our nations have the world's most sophisticated intelligence gathering capabilities. We are masters of electronic intelligence, of human intelligence, of signal intelligence, open-source intelligence, and technical intelligence. Yet for all that intelligence, it is a truism that the Arab Spring passed us by; the despicable attacks of September 11, 2001, came as a surprise and the emergence of the Islamic States in Iraq and Syria (ISIS) was not predicted. In this monograph, one of the world's leading experts in Information Operations explains the science behind what he calls population intelligence (POPINT). He explains how sophisticated social science research and behavioral profiling can be used to warn us of impeding issues, and how that information might be used by senior strategy makers as a tool for testing and refining strategy. This is not some ethereal dream; Dr. Tatham shows us that these techniques have been used already to great success. Yet he argues, forcefully, that we are collectively still caught up in old ideas and thinking. He makes a compelling case that the science of Target Audience Analysis (TAA) is now so well advanced that it must become a key component of future strategic decisionmaking.

In our current campaigns against ISIS, considerable resources have been ploughed into social media. Dr. Tatham argues this is a mistake. He views social media not as a precursor to behavior, but simply as just another communication conduit. He sees this as a continuum of wrong activities being undertaken. In Iraq and Afghanistan, he saw how big public relations and marketing companies cost the U.S. taxpayer millions of dollars in ultimately failed communication and propaganda campaigns. Social media, he argues, has become yet another blank checkbook for companies who rely on creative energy rather than empirical understanding to produce communications campaigns. Instead, he argues for far greater resource in TAA and greater understanding by federal agencies of what is and is not possible or desirable in their communication efforts. To this end, he looks in particular at U.S. Agency for International Development relief work in Pakistan and argues that the communication objectives set at the start of their projects are almost unattainable, even naïve, in their presumptions.

This is a rich and insightful paper from a previous Strategic Studies Institute author whose first paper with us in 2013, *U.S. Governmental Information Operations and Strategic Communications: A Discredited Tool or User Failure? Implications for Future Conflict*, created a storm of discussion and constructive debate. This latest paper is destined to follow the same trajectory.

DOUGLAS C. LOVELACE, JR.
Director
Strategic Studies Institute and
U.S. Army War College Press

ABOUT THE AUTHOR

STEVE TATHAM is the Director of Operations at IOTA-Global Ltd, a specialist civilian Information Operations company and is the contracted subject matter expert for Strategic Communication and Target Audience Analysis at the North Atlantic Treaty Organization, Centre of Excellence for Strategic Communication in Latvia. He resigned from the United Kingdom's (UK) Armed Forces in July 2014.

On leaving, he was the UK's longest continuously serving Officer in Information Activities. Between 1998 and 2003, he worked in Media Operations, covering conflicts in Sierra Leone (2000), Afghanistan (2001), and Iraq (2003), where he was public spokesman for the invasion. Between 2003 and 2013, he worked in Information Operations and Psychological Operations. Dr. Tatham was the Commanding Officer of 15 (UK) Psychological Operations Group from 2010 to 2013, during which time he deployed on multiple occasions to Afghanistan; was involved at the operational level in operations in Libya and has deployed to East Africa in an advisory role. His final military appointment was as the Special Information Operations project officer in the UK Ministry of Defence Operations Directorate. In 2007, he advised the then commander of British Forces in Afghanistan on influence operations when the strategically vital town of Musa Qala was retaken by British and Afghan forces. The Pentagon later described that operation as the "single best thing to come out of Afghanistan."

Dr. Tatham is the author of two books: *Losing Arab Hearts & Minds. The Coalition, Al-Jazeera and Muslim Public Opinion* (Hurst & Co, 2006) and *Behavioural Conflict: Why Understanding People's*

Motivations Will Prove Decisive In Future Conflict (Military Studies Press, 2012). Dr. Tatham holds an M.Phil. and a Ph.D. in international relations, both focusing on ideas of influence and strategic communication in conflict areas.

SUMMARY

This monograph revealed several recommendations for policymakers.

- Former Union of Soviet Socialist Republics leader Mikhail Gorbachev declared in November 2014 that: "the world is on the brink of a new cold war."[1] The West, he declared, had failed because of triumphalism—the belief that western, liberal, democratic free market economies were the answer to all ills. Yet across Eastern Europe, the Maghreb, the Middle East, and on the continent of Africa, unpredicted behaviors—some state initiated, some society initiated—have demonstrated a massive strategic deficit and an unpreparedness for "black swans."[2] The first challenge for policymakers is to accept that existing policy structures have not met the challenges of an increasingly interconnected and complex world; all too often the mechanisms to meet challenges are obscured by politics, process, received wisdom, complacency and fear of change. Our structures need to adapt: watchwords must be agility, risk, adaptation, innovation, and delegation to the lowest possible level and to the highest possible discomfort.
- The annexation of Crimea has added a new contender to ideas of existential threat; we have been slow to learn we are not well equipped to understand extreme Islam—we must use Crimea as an indicator of the potential for other threats—not necessarily motivated by religious ideology—and we must avoid complacency in assuming we understand why these events

occur. At the moment, one strand of North Atlantic Treaty Organization (NATO) strategy is the placement of troops in Baltic nations; how does this play with the organic Russian-speaking population in Latvia, Lithuania, and Estonia, for example. It may reassure the Latvian government, but does it make internal and external Russian operations more or less likely?

- Behavioral profiling is not just the preserve of police enforcement and Hollywood. The methodologies used to determine future latent group behavior are robust and have been verified and validated by defense science organizations in the United States and the United Kingdom; they could provide an excellent way to not only predict behaviors, but also to model response options in advance.
- Target Audience Analysis (TAA) can often provide very counterintuitive information, challenging existing received wisdom. For example, TAA may help understand if the Agency for International Development (AID) is effective in reducing terrorism—or as some have argued actually counterproductive. The answer to this is, of course, likely to be very country or region specific. Thus a proper TAA sweep would facilitate strategic understanding of where AID can be used as part of an effective public diplomacy project and where it should not.

ENDNOTES - SUMMARY

1. *Ex-USSR Leader Gorbachev: World on Brink of New Cold War*, BBC News, November 8, 2014, available from *www.bbc.co.uk/news/world-europe-29966852*, accessed May 3, 2015.

2. The theory of Black Swans was developed by Nassim Nicholas Taleb to explain the disproportionate role of high-profile, hard-to-predict, and rare events that are beyond the realm of normal expectations in history, science, finance, and technology; the noncomputability of the probability of the consequential rare events using scientific methods (owing to the very nature of small probabilities), and the psychological biases that make people individually and collectively blind to uncertainty and unaware of the massive role of the rare event in historical affairs.

USING TARGET AUDIENCE ANALYSIS TO AID STRATEGIC LEVEL DECISIONMAKING

INTRODUCTION

As the U.S.-led coalition comes to the end of its campaign in Afghanistan and perhaps contemplates the possibility of another protracted engagement dealing with the threats posed by the Islamic States in Iraq and Syria (ISIS), the wisdom of hindsight seems to flood bookshelves, staff college papers, and retiring commanders' valedictory media interviews. In retrospect, it appears that everyone knew what should have been done if only resource, time, political expediency, military capability—pick your own explanation—had allowed in the post-September 11, 2001 (9/11) years. Thus, we hear from the former Chief of the United Kingdom (UK) General Staff, General Sir Peter Wall, who told the BBC that, in Afghanistan: "We [the UK military] had put forward a plan saying that for the limited objectives that we had set ourselves, this [British Forces levels] was a reasonable force. And I freely admit now that calculus was wrong."[1] In considering the threat posed by ISIS, the U.S. President declared that: "we [the U.S. intelligence community] seriously underestimated the enemy";[2] while a U.S. Joint Staff report declared that:

> There was a failure to recognize, acknowledge, and accurately define the environment in which conflicts occurred, leading to a mismatch between forces, capabilities, missions, and goals . . . as a result, U.S. military training, policies, doctrine, and equipment were ill-suited to the tasks that troops actually faced in Iraq and Afghanistan.[3]

At the operational and tactical level, successive post operational tour reports bemoan the absence of this or that capability.

Of course, such reflection is not unreasonable, but what it perhaps surprisingly illustrates is that, despite the massive military and economic might of nations such as the United States and despite huge technological advantages, the conduct of warfare can still be pretty "hit and miss." The 9/11 intelligence community failings are well known, yet despite very extensive work put in place after that horrific event to rectify those failings and streamline our corporate understanding of the world, no intelligence service predicted the Arab Spring. Certainly, there were daily indicators that all was not well but not that various isolated events should combine into true regional revolutions—that was never seriously considered by anyone, let alone the huge internal security apparatus' of countries such as Egypt, Libya, and Syria. In 2002-03, the West was apparently certain that Saddam Hussein maintained his weapons of mass destruction (WMD) arsenals. Colin Powell told the United Nations (UN) that:

> Saddam Hussein and his regime have made no effort . . . to disarm as required by the international community. Indeed, the facts and Iraq's behavior show that Saddam Hussein and his regime are concealing their efforts to produce more weapons of mass destruction.[4]

Eleven years after that invasion, we know that was simply not correct.

In short, then, what we have seen over the last few years is widespread and senior recognition that huge swathes of governmental policy—be it U.S. or UK—have been misguided or ill-informed for a variety of reasons and that, despite the complexity of our collective intelligence networks, we still consistently fail to understand the environments in which we find ourselves working and the opposition that we face. But what if that wisdom of hindsight **could** have existed before the event? What if our intelligence mechanisms enabled us genuinely to understand the motivations and likely behaviors of, for example, the Taliban in advance, in a way that would have allowed us to model different strategies—to "road test" good ideas before their deployment. What if we had known what the stimuli would have been for the Al-Anbar uprising and we had been able to trigger it earlier? What if, armed with that knowledge, we could use it to trigger a rising against Islamic States (IS) occupation? And what of ISIS itself? We learned, finally, in Afghanistan that we should stop regarding the Taliban as a homogenous and cohesive group and instead focus on the disparate reasons for people attaching themselves to its aims. How useful would a similar understanding be for deciding strategy for dealing with ISIS?

Perhaps the rather surprising answer to all of these "What ifs" is that such a capability exists; it has existed for years, yet just like Sir Frank Whittle's invention of a jet engine in 1929, it is ignored by policymakers and advisers who seem unable or unwilling to embrace new ideas and who inexplicably, given the complexity of challenge today, remain wedded to old methods that cost enormous sums and achieve very little.[5] Take, for example, an invitation to tender issued by the U.S. Agency for International Development (USAID) in September 2014.[6] The objectives for a communication's plan in Pakistan list:

> The Pakistan Outreach and Communication Activity will implement a media and communication campaign to increase positive public perception and public support for the United States Government [USG] based on the development results achieved through the foreign assistance programming delivered by the [USAID] Mission to the Islamic Republic of Pakistan. This Activity continues previous efforts to move Pakistanis along the continuum from raising awareness about USAID/Pakistan's development projects to changing attitudes and perceptions about USAID, the USG, and the U.S. generally.[7]

These are the types of contracts that are embedded in the United States Agency for International Development (USAID), Department of Defense (DoD), and Department of State (DoS) programs. Their instigators believe that these public relations based programs actually can change the audience's minds favorably towards the United States. Yet, all the evidence would seem to suggest that this type of program is destined to fail. Why?

Four good reasons are: First, advertising U.S. largess is seen as boastful and locals find it insulting, demeaning; and patronizing. Second, the USAID brand, and in particular its logo, has already become synonymous with U.S. "interference" in local matters and, in the Muslim world at least, it is extremely unpopular. Third, the locals know perfectly well from where the money comes, and they are actually grateful, but do not want to feel like victims all the time. They are embarrassed by the incapacity of their own governments to provide for them, and they are unable to split the presence of USAID (which they appreciate) from wider U.S. foreign policy (which they do not). They might be grateful for the aid, but they still hate U.S. policy in the Arab and Muslim worlds and nothing, apart from a substantive policy change, is probably going to alter that view—certainly not a few sacks of grain or rice and an accompanying public relations campaign. Finally, any communications campaign will need the local organic media to be involved and, as soon as the local media starts publicizing U.S. generosity, it starts to lose its credibility in the minds of the locals. In short, the strategy is wrong.

Disagree? This author is not a fan of opinion polling per se, but trend analysis is useful when it is conducted empirically and by respected organizations. Take the much used Pew Global Attitudes Survey findings in Pakistan: "Roughly three-in-four Pakistanis (74%) consider the U.S. an enemy, up from 69% last year and 64% 3 years ago." Negative opinion in Pakistan towards the United States is rising, not dropping,[8] and over the last 12 years, it has remained consistently low—never better than 73 percent opposition to the United States; and in 2013, 89 percent opposition; this, despite massive public diplomacy and aid programs and accompanying communication campaigns.[9]

However, there is a dichotomy here. In the tender documentation, USAID acknowledges difficulties: "In addition, on-going regional, political, and social developments and public perceptions about U.S. policies have increased animosity in Pakistan towards the U.S."[10] But later it also declares that:

> the percentage of respondents who think "U.S. aid helps the Pakistani people," increasing from just under 40 percent in 2012, to over 50 percent in 2013. Polling also finds that respondents who have greater awareness of U.S. assistance to Pakistan hold more favorable views toward the United States and its programs.[11]

This is the problem with polls—they are highly temporal, depending upon the questions that are asked and, as we know from personal experience, what people say is often at odds with what they do. Thus to "square this circle," it is perhaps more insightful to look at actual behaviors. Here the picture is not as bright as USAID's tender document might have us believe.

The DoS travel advice portal states that travel to Pakistan should be avoided. It states that the U.S. Government continues to receive information that terrorist groups in South Asia may also be planning attacks in the region, possibly against U.S. Government facilities, U.S. citizens, or U.S. interests. The presence of al-Qaeda, Taliban elements, Lashkar-e-Tayyiba, indigenous sectarian groups, and other terror organizations, many of which are on the U.S. Government's list of designated Foreign Terrorist Organizations, poses a potential danger to U.S. citizens in the region. Terrorists and their sympathizers have demonstrated their willingness and ability to attack locations where U.S. citizens or Westerners are known to congregate or visit. The presence of several foreign and indigenous terrorist groups poses a danger to U.S. citizens throughout Pakistan. Across the country, terrorist attacks frequently occur against civilian, government, and foreign targets. Attacks have included armed assaults on heavily guarded sites, including Pakistani military installations and airports. The Government of Pakistan maintains heightened security measures, particularly in the major cities. Terrorists and criminal groups regularly resort to kidnapping for ransom.[12] The UK's Foreign and Commonwealth Office (FCO) seems to concur:

> The Foreign and Commonwealth Office advise against all travel to: the Federally Administered Tribal Areas; the districts of Charsadda, Kohat, Tank, Bannu, Lakki, Dera Ismail Khan, Swat, Buner, and Lower Dir in Khyber-Pakhtunkhwa; the city of Peshawar and districts south of the city, including travel on the Peshawar to Chitral road via the Lowari Pass; northern and western Balochistan; travel on the Karakoram Highway between Islamabad and Gilgit. The FCO advise against all but essential travel to: the Kalesh Valley, the Bamoboret Valley and Arandu District to the south and west of Chitral in Khyber-Pakhtunkhwa; the city of Quetta; the city of Nawabshah in Sindh Province, and areas of interior Sindh to the north of Nawabshah; Gilgit-Baltistan. There is a high threat from terrorism, kidnapping, and sectarian violence throughout Pakistan.[13]

In 2013, DoS warned Americans not to travel to Pakistan and ordered nonessential government personnel to leave the U.S. Consulate in Lahore because of a specific threat to that diplomatic mission.[14] In 2009, a U.S. journal investigated the most dangerous nations in the world to be an American citizen:

> We looked at the current geopolitical situation and a number of statistical categories, including rates of criminal offenses and deaths of U.S. citizens abroad, to come up with a list of the 12 most dangerous countries for Americans to visit.

Pakistan was number 12.[15] In May 2013, Reuters reported on a five-fold increase in unaccompanied U.S. diplomatic postings because of the danger to U.S. citizens' lives—Pakistan was one of five of the most dangerous posts listed.[16] In a September 2014 article, *The Diplomat Journal* predicted a deteriorating, not improving, security situation in Pakistan:

> For the time being, no group in Pakistan seems able to replicate the successes achieved by IS leader Abu Bakr al-Baghdadi in Syria and Iraq, but there is no doubt that the rise of the IS has galvanized the entire jihadist spectrum, making many groups more ambitious and aggressive than ever, and thus strengthening their attractiveness among the younger members of the Muslim population, which is a serious threat to the entire international community.[17]

Confusing? This is the problem of polling—polls are just not accurate predictors of real behavior and can provide very misplaced assurance. Neither is there a conclusive body of evidence to support the proposition that international aid efforts are effective counterterrorism tools—in fact, there is a substantial body of evidence to suggest just the opposite. The Brookings Institution has written that:

> Given aid's mixed record in supporting economic development and reducing poverty, what can we expect from aid in the fight against terrorism? The answer, in the short term, is not much.[18]

Professors Bu Savun and John Hayes of the University of Pittsburgh wrote in their 2011 paper: "aid is counterproductive if the recipient government uses repressive counterterrorism measures."[19] Even more disturbingly, some academics have argued that international aid can actually encourage terrorism, not reduce it:

> [M]any U.S. policymakers encouraged the use of foreign aid as an instrument to decrease terrorism. However, this argument is based on at least two as-

> sumptions that are the subject of considerable debate in the scholarly literature. The first is that terrorism is directly caused by problematic social or political conditions, such as poverty, lack of education, or oppressive governments. The second is that foreign aid is effective at alleviating any of those conditions, including terrorism directly. We contend that foreign aid is unlikely to have a pacifying effect on terrorism, and instead it might actually offer an incentive for the continuation or increase of terrorist activity.[20]

Given these competing tensions, and notwithstanding the morality of the world's richest nations supporting the world's poorest, is the publicizing of USAID really a sensible strategy? This author would suggest that it may not be and indeed has offered numerous previous examples of similar problems including in the December 2013 Strategic Studies Institute monograph of how a television campaign to quell violence against the United States in Pakistan had actually seen an increase in violence immediately afterwards.[21] Following the publication of that document, what was stunning was the significant numbers of U.S. military and diplomatic figures—some surprisingly senior—who made private contact with this author to make known their agreement and offer their support for the monograph's recommendations. From their vantage points in embassies and bureaus abroad, they offered a view that these types of campaigns actually worked very poorly.

UNDERSTANDING THE BASICS OF UNDERSTANDING

First, and surprisingly simple, all layers of authority should understand a single and very basic concept of communication which is this: Communication is two way—and what is received by the audience (which may not be the intended one) is often at significant variance to what is intended. This can be shown by two simple models that have been used in countless papers and books in the past to illustrate to nonsocial scientists and to nonspecialist communicators the mistakes that are being made and which are unashamedly repeated here.

In the 1950s, a model of communication was developed which became known as the "message influence model." This model was based upon some work undertaken by two academics, Claude Shannon and Warren Weaver. In 1949, they published their Mathematical Theory of Communication[22] which, at its inception, was designed to examine interference in telephone communication. It has subsequently—and perhaps erroneously (because it was never particularly bothered by the syntax of communication)—been used as a post-9/11 model for the failure of Western-led communication with global audiences. So, what does it look like and why is it relevant to us?

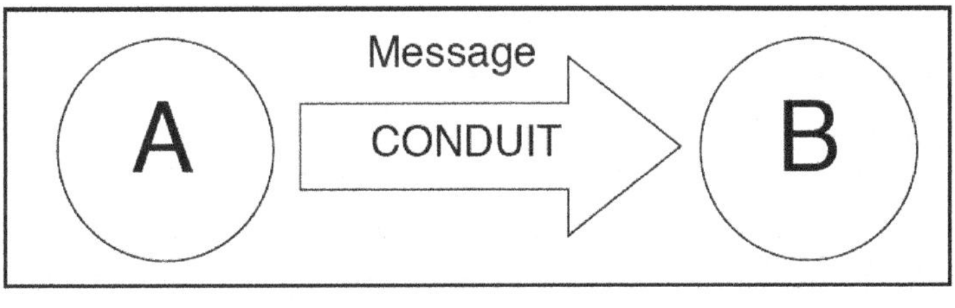

Source: Adapted from Shannon and Weaver, *The Mathematical Theory of Communication*, 1949.

Figure. 1. The Message-Influence Model of Communication.

Although simple enough to understand, the model has a number of flaws. First and foremost is that it simply does not reflect reality. Audiences do not sit passively and receive messages; instead they contextualize them according to a host of external factors. For example, in the case of Iraq, if we presume (A) was the United States and (B) the Iraqi people, the message of liberation and democracy may be heavily contextualized by elements of (B) in the light of revelations of torture by (A) at Abu Ghraib prison. Second, the model suggests that communication only occurs when a message actually is being transmitted, whereas actions and deeds (which are not modelled) send messages just as effectively or better than words—actions, as every schoolboy learns, speak louder than words. Third, the message presumes success—a potentially perilous supposition. This is perhaps best illustrated by a *Washington Post* article which records the attempts of U.S. Soldiers to combat stone-throwing youths in Iraq. The article reported that coalition forces in Sadr City, Baghdad, were facing a daily barrage of rocks thrown by young children. The problem for the coalition was how to stop it. Patently, violence, or even the threat of violence, against small children, was not an option, yet the stone throwing needed to end.

A U.S. Army psychological operations (MISO) team believed that they had an answer and crafted a series of leaflets which demanded that the children cease throwing stones. Yet, the leaflet campaign failed. Why? In this example, the messages to stop were interpreted by the children not as a warning, but as a sign of their success against the coalition. The message source was self-evidently "the enemy." The communications channel (the leaflets) did not resonate with the young children who either could not read or were not minded to read "adult" leaflets. Indeed, in this example, only the intended audience for the message was correctly identified by the MISO. Since the messages were received by the children without interference, the presumption, using the message influence model, was that the plan would be successful; clearly a more sophisticated model was needed.

As countless academic studies have shown, messages are heavily contextualized by outside events. So it seems "we" might need a new model on which to base "our" communication outreach. One possible solution is something called the Pragmatic Complexity Model, which draws on research undertaken by Niklas Luhmann.[23] Luhmann believed that communication was not the simple transmission of messages between two minds, but rather a complex system between sender and receiver. In any communication between party (A) and party (B), Luhmann believes that:

> The success of A's behaviour depends not only on external conditions but on what B thinks and does. And what B thinks and does is influenced by A's behaviour as well as B's expectations, interpretations and attributions with respect to A.[24]

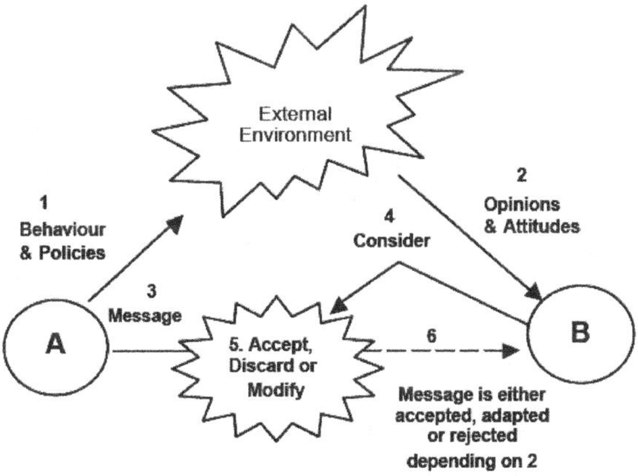

Figure 2. The Pragmatic Complexity Model (Author's Interpretation).

This model presumes that, in any communication, the success of (A)'s message depends not only on the message alone, but upon what (B) thinks and does. And what (B) thinks and does is influenced by (A)'s behavior and (B)'s expectations, interpretations, and attributions with respect to (A). The model assumes that messages are always interpreted within a larger and ongoing communication system, and that (A) and (B) are therefore locked into a relationship of simultaneous and mutual interdependence.

Here the success of (A)'s messages depend upon the wider external environment and, in particular, on (B)'s perception of (A)'s role in that environment. It is against that role that (A)'s messages are processed; they may be dismissed out of hand or they may be accepted, but in a contextualized manner. This model, which we think presents a much more realistic interpretation of society, suggests that there is no independent audience (B) waiting to be impacted by (A). Instead, both parties are locked into a relationship of interdependence.

Unfortunately, this model raises at least three further issues of complexity. The first is that the model presumes that (B) is passive. However, in reality (B) may itself be engaged in attempting to influence (A). Therefore (A)'s messages may themselves be contextualized by its perceptions of (B)'s actions. The second consideration is that if (A) can understand (B)'s opinions and attitudes in advance, (A) can prepare its messaging accordingly and thus attempt to mollify the effect of Step 5, thus creating a far stronger message. The third issue is identifying exactly who are (A) and (B). The names are overly simplistic, for they actually encompass many different, often disparate, groups, and these groups may themselves have some impact upon the message which may change it from that envisaged at sending.

If we apply this thinking to the USAID program, then the program will succeed or fail on parts 2, 4 and 5, i.e., the opinions, attitudes, and behaviors of the Pakistani aid recipients' (B), rather than on the U.S. Government's (A) message. Those opinions, attitudes, and behaviors are based on largely and uniformly hostile to the United States. As General David Patraeus told *Fortune* magazine in 2010, "We're not going to put lipstick on pigs."[25]

But that does mean that for the USAID program to work, very serious research has to be undertaken into the existing behavioral motivations to finesse the message and the conduit in a way that takes the audiences' behaviors into consideration and mitigates pejorative issues. This may reveal some counterintuitive issues—perhaps even that the United States is actually best serviced by **not** communicating that they have provided the aid—because that is the least worst option to pursue the desired end effect of reducing negative behaviors towards the United States by the Pakistani audience—or it may reveal some other unlikely conduit whereby the message can be sent. This idea of counterintuitive messaging is examined in later sections.

If you subscribe to these ideas, then it is obvious that the key step is understanding; the rest of this monograph will demonstrate how the proper employment of empirical social science methodologies can aid understanding of complex problems and assist strategy makers.

EMPLOYING SOCIAL SCIENCE FOR STRATEGIC DECISIONMAKING

In the early morning hours of January 16, 2013, a coordinated band of terrorists attacked a convoy of gas refinery workers as they departed the housing area of the In Amenas gas refinery in eastern Algeria. By the end of the 4-day siege that followed, at least 39 foreign hostages had been killed, along with 29 of the 32 hijackers. It could have been worse; some 700 Algerian workers and more than 100 foreigners managed to escape.

Following the attack, *The Times* newspaper carried a satirical cartoon showing the British Prime Minister playing whack-a-mole against a variety of international terrorist threats.[26] The cartoon referred to the need to react continuously to unexpected crises and threats. Since that cartoon was published, one could confidently add some extra holes: Ukraine, with a resurgent Russian President emerging; Iraq (with an ISIS fighter) and Nigeria (with a Boko Haram terrorist). Western politicians will continue to play whack-a-mole forever unless better efforts are made preemptively to understand the location and nature of as yet latent threats. For all the phenomenal intelligence architecture and power of the United States and its allies, it is clear that population intelligence remains elusive; it is still hard to predict what groups may do what to whom and why.

The solution to this problem is strategic target audience analysis (TAA)—the adaptation of a tactical psychological operations tool for strategic level problem solving. As we will see, TAA allows one to identify, in advance, key groups—who may not yet have emerged—through accurate behavioral profiling of groups. The process also provides a measure of "influenceability" of that groups—key information for policymakers to know in advance.

So let us return to USAID's problem. They want to distribute aid, and they want the Pakistani recipients to like, or at least not hate, the United States. Implicitly, they want to use this increased positive awareness of the "good" U.S. to reduce anti-U.S. behavior. This author has written extensively about the misplaced assurance that people have in the relationship between attitudes and behaviors, the fact that positive attitudes and perceptions are not strong precursors to good behavior but that, paradoxically, once a behavior has been established, there is a significantly increased chance that positive attitudes will follow. Therefore, this author would argue that USAID should first be either undertaking or placing contracts for social research of audiences, and that they should be considering their campaigns in a behavioral, not attitudinal, framework. Certainly, they should be looking at a program that will not just involve overpaid contractors broadcasting at them.[27] Any such program would involve a six-step solution:

Step 1: Identify what the audience was currently doing and thinking—and understand why. We might call this the Strategic Campaign Planning phase.

Step 2. Establish under what circumstances the audiences' behaviors could be changed and which triggers were the strongest—this is the TAA phase, and one of its key steps is to benchmark current behavior because that data is vital for Step 6 later.

Step 3. Armed with this information, present strategic options to the client balancing cost, risk, and likely effect.

Step 4. Design the strategic intervention—the Campaign Intervention Strategy.

Step 5. Deploy the campaign.

Step 6. Assess the resultant change in the audience—the Measures of Effect (MOE).

In its entirety, this is an empirical process and one that places the audience at the absolute center of the entire campaign. This might seem absolutely obvious, and yet it is astonishing how often the target audience disappears from strategy making. Consider this example from Afghanistan.

During the Winter 2007 deployment to Helmand, then Brigadier Andrew Mackay, the commander of British Forces, completely changed the lexicon of his 9,000 plus troops, affording primacy to the population rather than to dealing with the enemy. In his post operation tour report he wrote:

> In the [counterinsurgency] campaign, the population are the primary focus of all agencies, including the military, since this is a battle for consent and support. This is counter-intuitive to a military trained for war-fighting, whose natural inclination was to put the enemy at the center of its thinking.[28]

In essence, he asked his team to think of the conflict in Helmand as an eco-system—with undecided civilian population at its center, not the enemy, and the need to understand the behavior of that nodal point as the key driver for military operations. This was difficult, and Mackay freely admits that he was working not from doctrine or army manuals, but from behavioral economics books such as *Nudge* and *Predictably Irrational*. In both of these books, he saw solutions to operational military problems. That operational tour was characterized by success and has been described by many commentators as the tipping point in the battle for Helmand. Yet, in truth, Mackay's operational design pulled together different ideas and notions and attempted to place them into a military context fairly randomly and with no real guarantee of success at the outset. As it transpires, his agility of thinking and, more importantly, his absolute willingness to take risk, paid handsome dividends and has subsequently provided a real model for future conflicts. At the outset, the outcome was unknown, and, above all else, Mackay generated tremendous sense of purpose and leadership in ensuring his ideas were embedded among his sometime skeptical subordinates.

But what if the commander did not have to pull together randomly different ideas but instead was provided with a proper behavioral profile that he could have used in his military planning—one that modeled likely behaviors and offered, in advance, potential solutions. A distant aspiration or real possibility? Neither—for it is a reality, and between 2012-13, the behavioral profiling of group behavior was properly validated and verified by the UK Ministry of Defence (MoD). Farfetched? To some, for sure. But no more farfetched than the driverless car or internet enabled contact lenses—both recent Google inventions.

Behavioral Profiling.

Of course, behavioral profiling is not a new capability. In the West, behavioral profiling is now an established and recognized part of law enforcement operations. Take the case of the (metaphorical) serial killer; do they leave the eyes of their victims open? If they do, then police know they are searching for a particular type of personality—perhaps one that is vain and wishes to be admired and watched; the center of attention. Yet, if the perpetrator leaves the eyes shut, then the police have a completely different pool of potential perpetrators to search within—perhaps individuals who are shy, embarrassed, who hide away from attention and limelight. The Federal Bureau of Investigation's behavioral science unit, made famous by countless Hollywood films and now called The Behavioral Research and Instruction Unit, trains police officers from all of the world in this science of behavior. In the UK, Prime Minster David Cameron, soon after entering office, established the so-called "nudge unit," a behavioral science research branch inside The Cabinet Office, to look at how the emerging science of behavioral economics could be used in government applications—for example, in reducing rates of diabetes, and obesity, which place such a drain on the UK's publicly funded health service.

But this is very much in the realms of profiling after the event. What if you could profile in advance—to predict what actions groups—not individuals—may take? Many corporate organizations now employ psychometric testing to determine personality type and intellect—in order to see if the prospective job applicant will fit into the culture and ethos of the company in the future. The Revelian Behavioral Profile, for example, indicates how people are likely to approach problems, interact with others, and respond to the pace of the environment. Revelian's Cognitive Ability Test consists of 51 questions, which require applicants to process either verbal, numeric, or abstract information to answer the question correctly. The questions become more difficult as the test progresses and range in difficulty from those that 97 percent of people answer correctly, to questions that only 2 percent answer correctly. This broad range of question types and difficulty levels means that it is possible to assess general cognitive ability. Furthermore, by comparing individual's scores to relevant normative groups, the test can fairly accurately predict potential job performance. Of course, everyone is familiar with behavioral profiling for jobs and from the abundance of TV programs its application in criminal justice, but this is just the tip of the iceberg. For example, researchers studying animal behavior have always been aware that individual animals may differ in their behavior, for a number of reasons such as age, sex, physical condition, or past experience. An increasing body of evidence suggests that another factor responsible for consistent and (sometimes) predictable differences in behavior is temperament. Behavioral profiling is the term used to describe a variety of different methods that attempt to measure individual differences in temperament or personality to assist in the captive management of wild animals in zoos.[29]

In homeland security, behavioral profiling has become routine. As of March 2010, the Transportation Security Administration (TSA) deployed about 3,000 behavior detection officers (BDOs) at an annual cost of about $212 million; this force increased almost 15-fold between March 2007 and July 2009. BDOs have been selectively deployed to 161 of the 457 TSA-regulated airports in the United States at which passengers and their property are subject to TSA-mandated screening procedures.[30] In short, the use of social science to assess latent behavior is neither new nor revolutionary—and yet its application in the military domain remains corporately nugatory, despite attempts by individual commanders to shed light on this valuable tool.

This is surprising—not least because elements of the human terrain have taken such a notable focus of attention in recent operations. Biometric and DNA data collection, for example, enabled the mapping of human networks in the fight against improvised explosive device (IED) networks in Helmand. Yet, in this one fundamental area, there continues to remain a deep-seated weakness—that of understanding the human terrain as a collective whole from the perspective of behaviors. Observers of the Afghan conflict repeatedly have mentioned that there was a collective failure to understand the human environment in a meaningful manner. It is almost paradoxical that the more human terrain data was collected, seemingly the less was known and understood. Families, tribal linkages, and the identity and location of ideologues were known, but it proved difficult and often impossible to assemble all of this "noise" into meaningful and actionable data.

WHAT IS TARGET AUDIENCE ANALYSIS?

What TAA reveals about groups and societies can be quite surprising. For example, a major TAA project conducted for the U.S. Central Command at first sight appeared to support the notion that in Afghanistan the Taliban oppose education.[31] In fact, it showed that while some of the Taliban may, indeed, oppose education, this notion was actually too broad a generalization. In South Waziristan, for example, there were more schools under the Taliban than there are now after their removal. The TAA research revealed that the perception of Taliban opposition to education was the result not of religious zealotry, but actually of something far more common in every society, including the

West; have/have not rivalry between different socio-economic groups. Thus, if the result had been seeking to influence the local population against the Taliban by highlighting this as an endemic religious issue, the whole influence campaign would have failed because it was based upon a false premise. Similarly, the Kingdom of Saudi Arabia, home of the two holy cities of Mecca and Medina, has a reputation, in certain sections of the Western media at least, of being ultra-conservative. Women wear the veil and may not drive cars, and all other religions are banned. The country's justice code is infamously strict, with thieves having limbs amputated and adulterers being stoned. As a result, it is sometimes therefore assumed that Islam will be a dominant influence on young adult male behavior. Islam is important, yes. But TAA shows that it is simply not as important a behavioral trigger for young men as privacy from their parents, soccer, or nationalism. So an influence campaign aimed at this audience, grounded in Islamic references or context, might be less successful than one grounded in the fortunes of, say, the Manchester United Football Club.

Another common presumption is that, in Afghanistan, the tribe is the defining feature. This, too, is misleading. Despite the importance of tribes, empirical TAA data reveals that it is regions and land (and the incomes and status associated with them) that are considered to be much more important. So influence campaigns vested in tribal culture may well resonate, but may be less successful than ones focused on geography and land usage. Any reader unfamiliar with Afghanistan can consult a narrative of crossing the country alone on foot by British Member of Parliament Rory Stewart. In his book, *The Places In Between*, he describes how his guides refused to accompany him to the next village or valley, or beyond the next mountain, because it was someone else's territory.

TAA, when undertaken properly, is an extremely complex process, and often offers up completely counterintuitive findings. For example, between 1997 and 2000, there were huge increases in HIV transmission throughout the Caribbean. International funding was secured for a large public information campaign urging men to wear condoms. Yet, the slightly counterintuitive nature of a later TAA revealed that, in fact, men were simply not motivated to wear condoms at all—to do so was regarded by the target audience as a slur on their masculinity, and, in some cases, they believed that it would send a message that they were already diseased and therefore their chances of enjoying sexual relations would be much reduced. Thus the original target audience of young men was not actually the key audience at all—instead, the TAA revealed that young afro-Caribbean women were very worried about pregnancy, single-motherhood, and disease. Therefore, instead of an information campaign targeting men, what was needed was an unwanted pregnancy campaign targeted at women, as this would achieve the behavior change and arrest the HIV transmission. The campaign succeeded in introducing the use of condoms and significantly reducing HIV transmission rates. TAA, therefore, aims to construct a robust profile of the audience and how it can be influenced by an appropriately conceived and deployed message campaign. One key feature of this approach is that messages are developed in a bottom-up fashion, with them being constructed from a process of measurement and research, and subsequently derived from reliable knowledge of the audience. This is at odds with the current way that the military traditionally conducts its business, where themes and messaging are crafted centrally and distributed downwards to theatre troops. Political messages from Washington and other capitals are often a diluted and distant memory by the time they reach the tactical level, and they may actually have no relevance at ground level anyway.

Take the USAID program mentioned earlier—here the intent is to publicize the presence of USAID without any reference to the audience. At no stage does the invitation to tender invite research into the audience, it is entirely a "Push" function and, indeed, the tender—no doubt issued from Washington—is highly prescriptive. The contractor is to consider running events at community youth centers highlighting USAID's youth related activities; Ramadan-related community events; sponsorship of International Youth Day with extensive media and social media coverage; university workshops; Social Services Day in rural areas; Job fairs at universities; and events involving youth with special needs. The contractor will: develop numerous outreach materials, including talking points, scene setters, press releases, online materials, photos and short videos, as well as print materials, such as calendar and brochures. Scene setters, press releases, and other materials must conform to USAID/Pakistan's formatting standards and USAID's Branding and Marking guidelines. USAID/Pakistan produces and implements activities based on a mission-wide monthly events calendar. But there is no mention that the contractor will research the audience and determine what might actually be the best way to reduce hostile behavior. Instead, the pathway is fairly tightly articulated.

It should be clear why the TAA approach is far more effective than simple marketing approaches, or indeed even cultural understanding. There exists no universal model of communication applicable to all groups and cultures. All communication efforts must be tailored to the local dynamics and with respect to the behaviors one is seeking to change. Because audiences are multifaceted and cannot be grouped as a population, influencing the differing component groups of a society requires precisely targeted methods and approaches: One message—no matter how culturally relevant—does not fit all. Working out who to influence, why, how, when, and whether it is possible constitutes the first steps of TAA. Often, it will be necessary to influence one group to influence another. Above all else, the process of influence is not necessarily to make a particular group like "us" or "our" ideas—although this is always an extra bonus. It should be obvious, but as can be seen by the September 2014 tender—it is not.

There are some further issues with TAA that merit consideration. If we think of TAA as the process of identifying the "right" audience, we must also be mindful that there are other audiences also present. We can think of them in four distinct spheres: the target audience; a group who may react positively to the messaging applied to the target; a group who may react negatively to the messaging applied to the target; and a group who will be ambivalent and who might even be best left alone.

TAA IN ACTION

In the UK, at least, there has been a gradual realization that TAA is a key component in future operations. To try and understand TAA capabilities better, this monograph's author defined for the UK MoD three tiers of TAA capability. Tier 3 TAA is the least detailed TAA and is almost exclusively derived from secondary research. For example, this is typically remote, open source analysis of target groups, very often carried out in the language of the analyst rather than in that of the target. For example, this may be an Internet based research project on a specific group—Alawites in Syria, or Kurds in Iraq. Invariably, it will try to find third party studies, perhaps derived from academic or nongovernmental organizations (NGOs), and aggregate the information for military usage. This is usually open source, but it could also involve classified intelligence. This author explains the output of Tier 3 TAA as "assumed information."

Tier 2 TAA is any primary research involving actual contact with the audiences of interest but, critically, it does not follow a specific scientifically verified deductive methodology. It may be conducted in-country or remotely, and is largely attitudinally based. The output of Tier 2 TAA is information recorded from interactions with target audiences. An example of Tier 2 TAA is a report from a patrol or a *shura*, where soldiers ask locals what they think is going on and what actions might positively influence attitudes and behaviors. A refined variation might be a cultural advisor on patrol. This type of TAA may with time become quite detailed. It provides another layer of data over and above that of Tier 3.

By far the most useful TAA, however, is Tier 1. This is a multisource, scientifically verified, diagnostic methodology undertaken in-country and in the local language used to identify specific motivations for behavior. The output of Tier 1 TAA is information deduced from methodically gathered data, and tested against a scientifically derived hypothesis. To illustrate this, Figure 3, taken from an actual case study undertaken by the UK Government in 2013, shows the behavioral profiles of a specific target group—young unmarried males—in the target country. In essence, it is a simple graphical representation of the key components that will influence a target group's behavior. They are derived from a broader set of research parameters shown in Figure 4, and it is of note that of 29 different components that are researched and evaluated to create the behavioral profile grid, only one looks at current perceptions and attitudes (which forms the bulk of most other non-TAA derived communication programs).

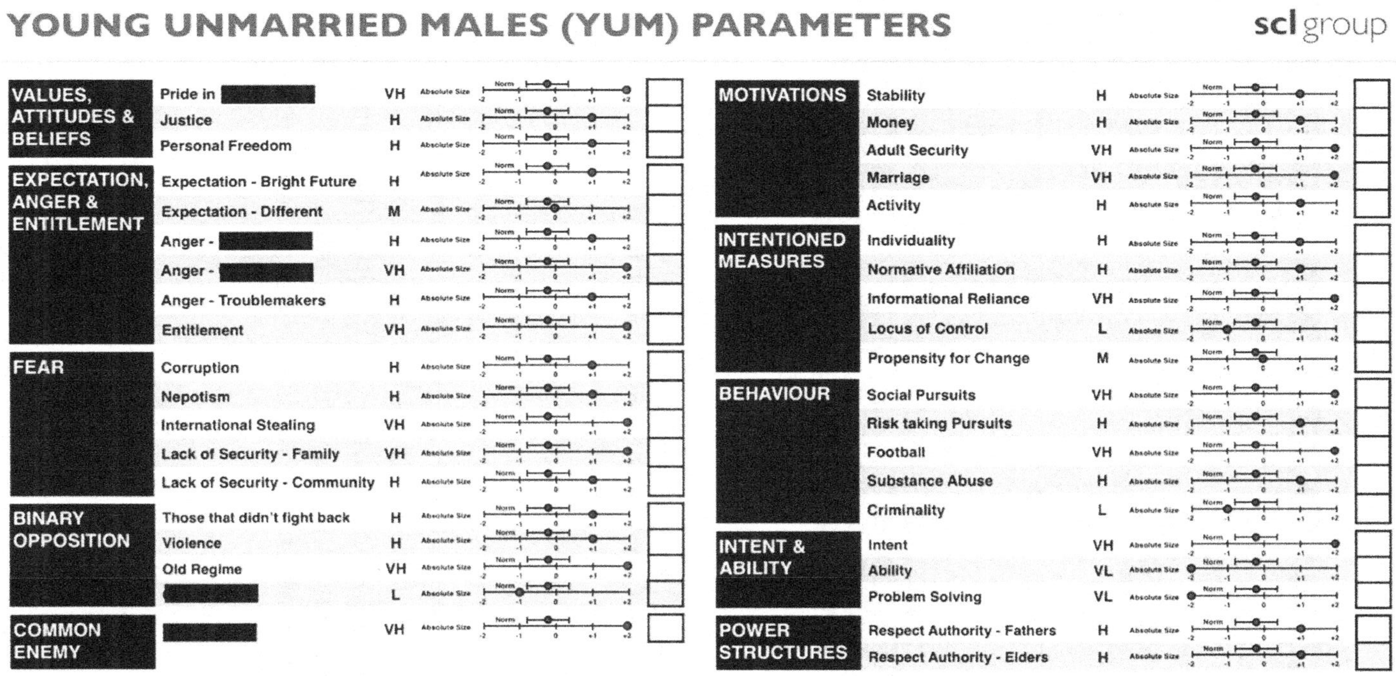

Figure 3. The behavioral Parameter Scale for the Target Audience of Young Unmarried Males.

DESCRIPTIVE	PROGNOSTIC	TRANSFORMATIVE
AUDIENCE STATISTICS	INFORMATIONAL	MESSAGE APPEALS
LANGUAGE	PROPENSITY FOR CHANGE	COMMON ENEMIES
LITERACY	EMIC LOGIC	BINARY OPPOSITION
CURRENT BEHAVIOUR	NORMATIVE AFFILIATION	SOURCE CREDIBILITY
SIZE	LOCUS OF CONTROL	REWARD STRUCTURES
LEADERSHIP	MOTIVATIONS	DECISION PATH
FINANCING	POLARISING MOTIVATING PROPENSITY	FILTERS
RELEVANT ISSUES		NOISE
HISTORIC CONTEXT	INSTRUMENTAL	ATTITUDES
CHANNEL CREDIBILITY	INFLUENCEABILITY	INITIATING SETS
GROUP COMPOSITION	BELIEFS	RITUALS
MISSION	INTENT	ABILITY
VALUES	SKILL	EXPECTANCY
GROUP MEMBERSHIP	POWER STRUCTURES	FEARS
CHANNEL EXPOSURE	IDEOLOGY	
NEEDS	FRAGILITY / RESISTENCE	

Figure 4. Overview of Research Sets for the Creation of Behavioral Profiles.

A detailed examination of each of these criteria is beyond the scope of this monograph, but a cursory examination of just two is illustrative as it shows the depth of research that is required to make detailed strategy recommendations. The two examples are Locus of Control and Source Credibility—the former an internal psychological attribute the latter a key determinate in reaching an audience with a message.

Locus of Control.

A Target audience's locus of control represents that audience's view of what determines the course of future events. A key distinction that must be determined are those groups who can be characterized by an external locus of control, for whom control over events is believed to be external to them, and those characterized by an internal locus of control, who believe that control of external events is located in one's own actions. A number of tests exist to measure locus of control.[32] A number of revelations flow from the finding that a given audience is characterized by, for example, an external locus of control. Such an audience places a lower value on incoming information, because such information is believed to have little effect on external events. An audience with an external locus of control is less inclined to take action, because in their worldview, doing so will have little influence on how matters will play out. By contrast, an audience with an internal locus of control is inclined to take stock of a situation and assess how a different course of action might produce more desirable results. A willingness to await delayed gratification is associated with an internal locus of control, as are a tendency to resist coercion and a proclivity to take risks. An extreme finding for a target audience's locus of control is a particularly consequential result to emerge from audience analysis, and one that can play a major, even definitive role in the construction of an effective strategic communication campaign. Assessing a target audience's locus of control is essential to understanding what strategic communication has the capacity to accomplish vis-à-vis that audience, as an external locus suggests that a revision of behavior is unlikely whereas an internal locus indicates greater potential. Moreover, determining an audience's locus of control can enable the crafting of messages that properly, effectively, and convincingly assign causes of past occurrences and suggest potential reasons for future developments. For example, historically, the Nigerian elections have been notoriously corrupt, with blatant fraudulent practices being witnessed by international observers and the media. A research project of the electorate to establish under what conditions the public might rise up against the government if fraudulent practices were observed actually revealed that there were almost none—because their locus of control was particularly low.

Source Credibility.

This is a measure of the trustworthiness of a particular source (or origin) in the eyes of the relevant target audience. Just as messages can have differing receptions and impacts depending on the channels through which they are broadcast, so too can messages differ in their influence depending on the sources by which they are sent. The relative credibility of various sources within a given target audience can be measured through questionnaire and as a complement to such quantitative methods, surveys and focus groups can offer qualitative data for understanding the credibility of a particular source in the eyes of the target audience. The credibility of a particular country as a source for communications can differ over time, in different places, and among different age groups. Moreover, the country's credibility can differ depending on the topic or issue being addressed by a strategic communication campaign. It is not enough for a message to reach its audience; that message must be accepted by, and thus influence the behavior of, the target audience. Hence, the source associated with the message is crucial; and, in particular, the credibility of that source will have major consequences on the extent to which the message is, in fact, accepted. Psychological experiments have underscored the importance of considering source credibility in crafting strategic communication. Psychologists, beginning with Carl Hovland, Irving Janis, and Harold Kelley, have noted that messages likely to appear sensible to an audience are generally accepted regardless of the source. For messages whose content is more questionable, however—as is the case with most strategic communication—only those messages emerging from a source considered credible will be accepted by the audience. Experiments have shown that, because of the enormous exposure to commercial marketing campaigns, even less developed communities are aware that a "commercial" or "advertisement" is an appeal by an interested party (the communicator) to try to influence your attitude or behavior. Therefore, appeals without any apparent source are deeply distrusted. Many of the "grey" programs that were run in Iraq had no identifiable source and therefore lacked credibility with their intended audiences.

Having profiled the audience's behavioral characteristics a possible strategy is modeled against the profile—in this instance, the strategy is some form of national duty. In Figure 5, it can be seen by the preponderance of green lights that this is a good strategy and would work with the target group for the desired behavioral outcome.

However, in the instance that follows, which is the giving of financial aid, it can be seen that the profile is populated by red lights, indicating that this is not a good strategy to be employed, see Figure 6.

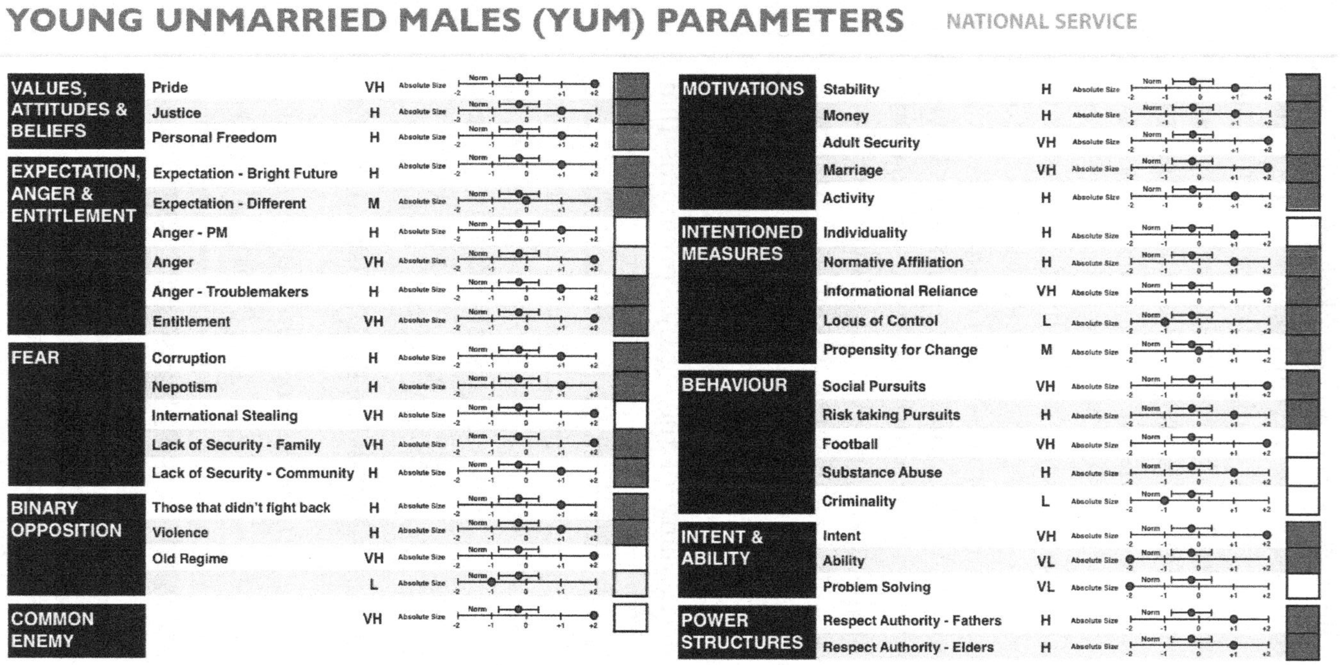

Figure 5. The Behavioral Parameter Scale Populated for the Impact of National Duty on the Target Audience of Young Unmarried Males.

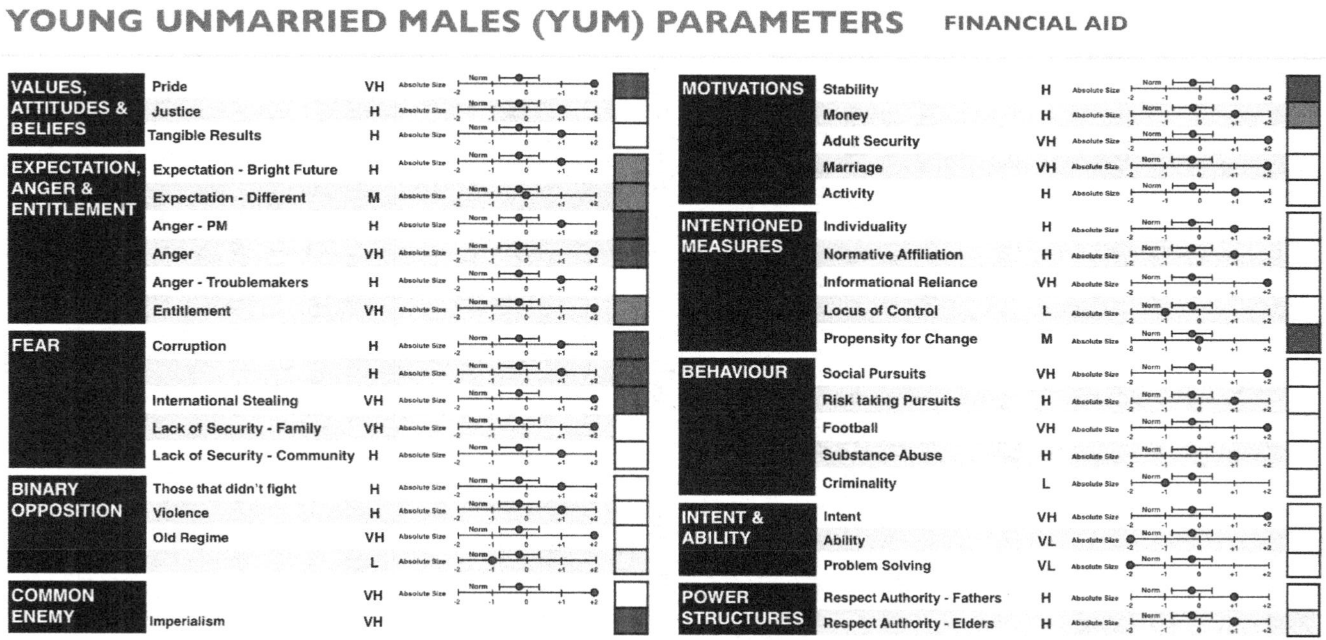

Figure 6. The Behavioral Parameter Scale Populated for the Impact of Financial Aid on the Target Audience of Young Unmarried Males.

As was illustrated with various examples previously, TAA has been used very successfully in a series of applications and geographical regions. Increasingly, however, a new field of research has opened up that superficially appears to have potential to compete with TAA as an indicator of future behavior—social media analysis.

DOWN THE SOCIAL MEDIA RABBIT HOLE?

The wave of protests that swept the Arab world between 2010 and 2014, and which in many instances have still to reach their conclusion, if nothing else seem to have become the benchmark by which the impact of social media is assessed, at least as it pertains to the study of international relations and conflict. Social media was a key facilitator for political change; the Project on Information Technology and Political Islam found that in the week before Hosni Mubarak stepped down, the number of tweets in Egypt and around the world about the political developments in the country jumped from 2,300 a day to 230,000 a day.[33] Of course, how many of those emanated from within Egypt is a different issue. The social media intelligence firm Sysomos analyzed 52 million Twitter users during the revolutions in Egypt, Tunisia, and Yemen, and concluded that only 14,642, or 0.027 percent, positively identified their location as Egypt, Yemen, or Tunisia.[34] This figure is low even when allowance is made for the fact that only a small proportion of Twitter's approximately 500 million active users and their 72 billion tweets (figures from 2013) opt-in to allow Twitter to broadcast their location with every tweet.

Even if location is disabled, it is possible to analyze the metadata associated with each tweet, such as time zone and language, to search for specific trends. Data generated by Twitter users and available through Twitter's application programming interface by one study found that during the 1-week sampling period, roughly 20 percent of the tweets collected showed the user's location to an accuracy of street level or better.[35] This is important to remember because in the last few years, the analysis and interpretation of social media has become a growth industry, with numerous companies now suggesting their algorithms will allow accurate prediction of all type of behaviors—from purchasing inclinations to propensity to cause violent rebellion. This commercial certainty is not mirrored in academic study. "It is unclear if and how social data can predict behavior, and whether such predictions are more accurate than those arising from current marketing practices," assert Sharad Goel and Daniel Goldstein in their study, *Predicting Individual Behavior with Social Networks*.[36] The predictive value of social media has "yet to be established in any meaningful way," according to psychologist, Dr. Jane Adams.[37]

Haile Owusu, director of research at SocialFlow, studied social media data following the fatal shooting of 17-year-old Trayvon Martin by George Michael Zimmerman in Sanford, FL, in February 2012. He concluded that there was:

> a violent tone to a lot of what is written about Trayvon and Zimmerman that hasn't precipitated anything. . . . Saying is very different from doing, and social media is often used as a place to vent, and nothing more. Twitter cannot predict if or when violent words will become violent actions.[38]

In their study, *Real-World Behavior Analysis through a Social Media Lens*, Mohammad-Ali Abbasi, Sun-Ki

Chaiz, Huan Liuy, and Kiran Sagooz suggest that it is far easier to spot retrospective patterns and clues in Social Media when you know what you are looking for:

> Despite years of trying, no one's quite figured out how to harness the monster of publicly available online data to predict the future. It's a formidable task—Twitter alone sees some 58 million tweets every day. Using social media to predict violent social uprisings is almost impossible.[39]

Some academics remain nonetheless optimistic. Nathan Kallus writes that:

> With public information becoming widely accessible and shared on today's web, greater insights are possible into crowd actions by citizens and non-state actors such as large protests and cyber activism. . . . The study validates the common intuition that data on social media (beyond mainstream news sources) are able to predict major events.[40]

Sitaram Asur researched:

> chatter from Twitter.com to forecast box-office revenues for movies. We show that a simple model built from the rate at which tweets are created about particular topics can outperform market-based predictors.[41]

The key question is whether social media analysis provides data that was previously unobtainable. Professor Kalev Leetaru of the University of Illinois, Urbana-Champaign, IL, suggests that:

> Despite being hailed as a social media revolution, monitoring the tone of only mainstream media around the world would have been enough to suggest the

> potential for unrest in Egypt. While such a surge in negativity about Egypt would not have automatically indicated that the government would be overthrown, it would at the very least have suggested to policy-makers and intelligence analysts that there was increased potential for unrest.[42]

Lawrence Pintak, author of *The New Arab Journalist*, pointed out on CNN that, despite the speed with which the Mubarak regime fell, bloggers and digital activists had been working toward reform under violent repression in the Middle East for years. "This is a digital revolution that has been happening for quite a while," he noted.[43] The director of the Council of the Advancement of Arab British Understanding wrote on Al-Rabiya's website:

> ISIS and like-minded groups are populated by those born in the Internet age, totally at ease with advanced programming and ICT skills. Their videos, their newsletters and use of social media are slick and professional. The trouble is that the real reasons ISIS and other extremist groups have been successful is only partially due to social media and far more to do with international and regional policy failures. Just as there was no Twitter revolution in Iran and no Facebook revolution in Egypt, ISIS is not merely a social media phenomenon. The real questions are why does ISIS attract followers and why does the West and its allies have such a poor record in countering this?[44]

This author would suggest that the answer to the last question is that because our collective communication campaigns lack detailed understanding at their inception.

Where there is perhaps more agreement is that social media acted as an accelerator to change. Eira Martens, a research associate with the Deutsche Welle

Academy, Bonn, Germany, showed that, in addition to helping organize protests, social media—in particular shared photos and videos—allowed participants to form a collective identity. This increased a sense of solidarity and helped lower a "fear threshold" that could otherwise have prevented people from taking to the streets.[45]

But the simple fact is that social media like Facebook and Twitter did not **cause** revolutions; the revolutions were caused by people, and for a variety of reasons. But social media did speed up the process by informing and by helping to organize the revolutionaries, and by transmitting their message to the world and galvanizing international support. A 2012 study undertaken by Sandia National Laboratories, Livermore, CA; Defense Advanced Research Projects Agency (DARPA), Washington, DC; and SCL Ltd., Washington, DC; determined that dispersion of discussion across European and Middle Eastern blog network communities did provide early clues of large mobilization events, but not of specific behaviors.[46] Sascha Meinrath, director of the New America Foundation's Open Technology Initiative, wrote that:

> Social media have become the pamphlets of the 21st century, a way that people who are frustrated with the status quo can organize themselves and coordinate protest, and in the case of Egypt, revolution.[47]

Rafat Ali, a social media expert and founder of PaidContent,[48] said Facebook and Twitter played different roles in the uprising. "Facebook helped to organize the activists inside the country," he said:

> while Twitter functioned to help get the message out to the broader world. Facebook acts like an accelerant to conditions which already exist in the country.

> Twitter and YouTube serve as amplification for what's happening on the ground. And they directly affect Western media coverage.[49]

What is clear is that, even in the era of big data, it is currently impossible to rely absolutely on social media analysis as a strategic tool for predicting latent events. For information on current events already occurring, social media remain a gold mine—but in a truly global information space, it will continue to be very difficult to determine what is relevant and what is not. Thus despite the "sexiness" of social media, strategic TAA remains the best and quickest way to determine latent behaviors.

HOW WOULD IT WORK?

Clearly, it would be impossible to TAA every group of interest in every country of the world. Quite aside from the vast expense, it would be largely nugatory work. Yet, in the West, there are around 10 to 15 countries that will endure as places of great interest—possibly because they are regarded as threats (past or present) to international security and stability; perhaps because they are emerging from conflict and their futures, and their future leaders, have not yet emerged; some because they are at risk of insurgency or internal corruption; and some nations because they are likely to fall into conflict and the West may be pulled into peacekeeping or stabilizing missions. The maintenance of an ongoing TAA research program on each country would be comparatively cheap, require a small footprint, could be covert or overt—i.e., it could be done with the full agreement of the host country—and could be continuously monitoring the evolution of popular groups and their behaviors. More importantly, as we have seen from the case study just discussed, it would allow desk officers for specific countries to run policy options and strategies against real and empirically derived data, providing political leadership with far more detained understanding of the likely consequence of their policy decisions.

Would it cost money? Of course, deploying research teams and crunching data to provide meaningful strategy recommendations is never going to be cheap. In 2013 when the UK MoD ran its trial, it was estimated that a 3-year TAA program in three countries could be procured for around £2 million; this might equate to less than £10 million for 10 nations for 3 years. This seems like a lot until you compare it with other defense expense. The UK spent £2 million for a public relations company to run communications for the Geneva II conference on 2014; one single Abrahams M1A2 tank costs over $6 million[50]; while the cost of deploying a single soldier to Afghanistan, for 1 year is assessed by Harvard's John F. Kennedy School of Government as being $1 million—the conflict having cost the U.S. Government the equivalent of $75,000 for every U.S. household.[51]

What is perhaps more troublesome is where this capability would reside in government. Here you have a very serious structural problem, addressed in following paragraphs, that TAA reveals whole of government solutions, and governments tend to work, despite efforts to the contrary, in departmental stove pipes. Thus, in the UK, the sense was that the capability needed to be either in the Cabinet Office, as the political center of gravity for government or in the National Security Council Secretariat.

WHY GOVERNMENTS WILL NOT INVEST IN STRATEGIC TAA

Professor Julian Lindley-French is Distinguished Visiting Research Fellow at the Institute for National Security Studies in Washington, DC, and is a former member of the UK Chief of Defence Staff's advisory panel. He is deeply pessimistic about the future:

> War is coming, big war. Not here, not now but some time, some place this century it is coming. The rapid shift in the military balance of power away from the democracies, arms races, climate change and the coming dislocation of societies, the dangerous proliferation of dangerous technologies, demographic pressures, competition for energy, food and water and the hollowing out of states. All the necessary ingredients for big war exist, driven daily by the growing systemic frictions apparent in the world.[52]

If we can learn one single lesson from Afghanistan, it is that conventional wisdom and a "can do" attitude were simply not enough to prevail. The adaption and innovation promised by senior officers now needs to be demonstrated and funded. Since the root cause of all conflict is people, understanding people better must be the starting point if we are to prevail in Lindley-French's "big war." Yet, there are significant obstacles in the way. Professor of History at the University of Chicago and President of the Adlai Stevenson Institute of International Affairs William Polk recently wrote in *The Atlantic Journal* that:

> As we have seen in each of our recent crises—Somalia, Mali, Libya, Syria, Iraq, the Ukraine and Iran—'practical' men of affairs want quick answers: they say in effect, 'don't bother us with talk about how we got

> here; this is where we are; so what do we do now?' The result, predictably, is a sort of nervous tick in the body politic: we lurch from one emergency to the next in an unending sequence. This is not new. We all have heard the quip: 'ready, fire, aim.' In fact those words were not just a joke. For centuries after infantry soldiers were given the rifle, they were ordered not to take the time to aim; rather, they were instructed just to point in the general direction of the enemy and fire. Their commanders believed that it was the mass impact, the 'broadside,' that won the day. Our leaders still believe it. They think that our 'shock and awe,' our marvelous technology measured in stealth bombers, drones, all-knowing intelligence, our massed and highly mobile troops and our money constitute a devastating broadside. All we have to do is to point in the right direction and shoot. So we shoot and then shoot again and again. We win each battle, but the battles keep happening. And to our chagrin, we don't seem to be winning the wars. By almost any criterion, we are less 'victorious' today than half a century ago.[53]

Iain Richardson is a former Royal Navy Captain with a distinguished career in military intelligence, and former Deputy Head of the now defunct UK Defence Academy's Advanced Research and Assessment Group. Based on his extensive experience of bringing future threats and problems to the notice of senior leadership, Richardson summed up the strategy deficit in a conversation with this monograph's author thus: "Our leaders are far too busy being busy to be anything other than busy."[54] In effect, process rather than strategy is the key deliverable in many departments of government throughout the Euro-Atlantic community, and, in keeping the process alive, strategic thinking inevitably takes second place. The result is that policy is almost exclusively reactive, as the author saw this first hand in nearly 2 years working on multinational problems while serving in the UK MoD. The pattern is repeated over and over; little or no resource is applied to understanding up-stream population-based threats, and, as issues develop in a particular region, staff who invariably have little understanding or knowledge of the country concerned are sucked into crisis teams to "firefight" issues. Those teams become under intense political and public pressure to deliver; resources are suddenly made available; the lure of public relations and marketing based campaigns which promise quick delivery becomes irresistible, and the more slow and considered process of researching and understanding is subjugated to the need to demonstrate delivery—measures of performance, rather than MOE.

During a trial of the strategic TAA approach described herein, few senior people had time or inclination to attend briefs, and, for those that did, while all declared that the data was astounding, no one was prepared to allocate increasing scarce financial resource on "nice to have" research in countries where there may or may not be a problem. In short, governments play the odds and when the bet fails, there is a substantial price to pay.

A third reason why strategic TAA seems likely not to be part of future government tools is that TAA is unconventional. It is like intelligence, but it is more than intelligence; it is like social science research, but it is more than social science research; it is like strategic communication, but strategic communication is but a part. It is related to defense, but also cuts across international relations and international development. All of this means that it does not fall naturally into one section or department of government, and thus officials need to take risk outside of their comfort zones. In some ways the current status of strategic TAA resembles the early development of Whittle's jet engine—both for the revolutionary effect that mainstream adoption would have on conflict, and for its tortuous journey towards acceptance.

As illustrated by the whack-a-mole example provided earlier, the emergence of new threats over the last few years has been relentless and wearying, but strategic TAA holds a key to pre-empting future problems in order to avoid costly and painful military engagements.

> The doctrine of 'exit strategy' fundamentally misunderstands the nature of war and the nature of historical action. The knowledge of the end is not given to us at the beginning.[55]
>
> Leon Wieseltier,
> American Philosopher and Writer

> . . . but with Strategic TAA, it could be.[56]
>
> Dr. Steve Tatham

ENDNOTES

1. "UK Military 'Made Wrong Calculations' on Afghanistan," BBC News, November 23, 2014, available from *www.bbc.co.uk/news/uk-29714738*, accessed May 3, 2015.

2. "Barak Obama: U.S. 'Underestimated' ISIS, Overestimated Iraqi Army," NBC News September 28, 2014, available from *www.nbcnews.com/storyline/isis-terror/obama-u-s-underestimated-isis-overestimated-iraqi-army-n213506*, accessed May 3, 2015.

3. *Decade of War*, Vol. I, "Enduring Lessons from the Past Decade of Operations," June 15, 2012, available from *cgsc.contentdm.oclc.org/cdm/ref/collection/p4013coll11/id/2035.*

4. "A Policy of Evasion and Deception," *The Washington Post*, February 5, 2003.

5. Frank Whittle, a British Royal Air Force officer, took a design for a jet engine to the British Government in 1929, but was turned down for funding on the grounds of impracticality. In 1930, he patented the design himself, but, having sunk all his funds into research, he could not afford the fee when the patent renewal was due in 1934. He again applied for government sponsorship, which again was declined. He raised £2,000 in private finance. In 1937, after 8 years of further research and development, he again offered the project to the British Government, which again declined to assist him. It was only in 1939 that a single government official, at risk to his career and reputation, backed Whittle and his project for funding. The result was that British jet aircraft only entered operational service at the end, not the beginning, of World War II.

6. Pakistan Outreach and Communication Activity Solicitation Number, REQ-391-14-000128.

7. *Ibid.*

8. "Pakistani Public Opinion Ever More Critical of U.S.," Pew Global Poll, June 27, 2012, available from *www.pewglobal.org/2012/06/27/pakistani-public-opinion-ever-more-critical-of-u-s/*.

9. *Ibid.*

10. Pakistan Outreach and Communication Activity Solicitation Number, REQ-391-14-000128.

11. *Ibid.*

12. "World Wide Caution Notice, Washington, DC: DoS, available from *travel.state.gov/content/passports/english/alertswarnings/worldwide-caution.html*, accessed January 13, 2015.

13. FCO Foreign Travel Advice, Pakistan, available from *https://www.gov.uk/foreign-travel-advice/pakistan*, accessed January 13, 2015.

14. "U.S. Issues Warning over Pakistan Threat," *USA Today*, August 9, 2013, available from *www.usatoday.com/story/news/nation/2013/08/08/us-lahore-pakistan-evacuation/2634337/*.

15. For the previous quote and this information, see "The 12 Most Dangerous Countries for Americans to Visit," *The Examiner*, available from *www.examiner.com/article/the-dangerous-dozen-the-12-most-dangerous-countries-for-americans-to-visit*, accessed December 14, 2015.

16. "Danger and Separation from Families Changing Job of U.S. Diplomats," Reuters, May 19, 2013, available from *www.reuters.com/article/2013/05/19/us-usa-diplomats-idUSBRE94I01G20130519*, accessed May 6, 2015.

17. "The State of Terrorism in Pakistan," *The Diplomat* September 8, 2014, available from *thediplomat.com/2014/09/the-state-of-terrorism-in-pakistan/*, accessed June 6, 2015.

18. *Can Foreign Aid Help Stop Terrorism?* Washington, DC: The Brookings Institution, Summer 2002, available from *www.brookings.edu/research/articles/2002/06/summer-terrorism-graham*, accessed June 3, 2015.

19. Bu Savun and John Hays, "Foreign Aid as a Counterterrorism Tool: Aid Delivery Channels, State Capacity, and NGOs, APSA 2011 Annual Meeting Paper, available from *ssrn.com/abstract=1900690*, accessed June 3, 2015.

20. S. Sharma, Z. Davis, and M. Findley, *Deadly Gift? The Effect of Foreign Aid on Terrorism*, Provo, UT: Brigham Young University, Department of Political Science, February 12, 2009.

21. Steven Tatham, *U.S. Governmental Information Operations and Strategic Communications: A Discredited Tool Or User Failure? Implications For Future Conflict*, Carlisle, PA: Strategic Studies Institute, U.S. Army War College, available from *www.strategicstudiesinstitute.army.mil/pubs/display.cfm?pubID=1182*.

22. C. Shannon and W. Weaver, *The Mathematical Theory of Communication*, Urbana-Champaign, IL: University of Illinois Press, 1998.

23. N. Luhmann, *Social Systems*, Stanford, CA: Stanford University Press, 1995.

24. *Ibid.*

25. "General Petraeus on Leadership (His Views on Truth)," *Fortune Magazine*, March 5, 2010, available from *archive.fortune.com/2010/03/04/news/companies/petraeus_leadership.fortune/index.htm*, accessed June 3, 2015.

26. See *www.englishblog.com/2013/01/cartoon-wac-a-terrorist-.html#.VW75aULRxl.*

27. In 2010, the Royal College of Defence Studies in London, UK, hosted a private seminar on behalf of then Chief of Defence Staff General Sir David Richards. At that event, to which numerous commercial companies had been invited, one contractor stood up and publicly stated that the loudest voice would always win through, and that repetition of message was utterly key to success. This author was stunned since no mention was made of the audiences—as if they did not matter and had no hand in the outcome.

28. Unclassified quote taken from 52 Bde, Commanders Post Operation Tour Report, April 2008.

29. Zoo Research Guidelines: "Behavioural Profiling," available from *winghamwildlifepark.co.uk/wp-content/uploads/2012/12/Behavioural-profiling-guidelines.pdf.*

30. "AVIATION SECURITY: Efforts to Validate TSA's Passenger Screening Behavior Detection Program Underway, but Opportunities Exist to Strengthen Validation and Address Operational Challenges," GAO Report 10-763, Washington, DC: U.S. Government Accountability Office, May 2010.

31. "Strategic Multi-layer Assessment for a Rich Contextual Understanding of PAKAF—Provisional."

32. These include Rotter's 23-item forced scale, the Crandall Intellectual Ascription of Responsibility Scale, the Nowicki-Strickland Scale, and the Duttweiler Internal Control Index.

33. Philip Howard *et al.*, "Opening Closed Regimes: What Was the Role of Social Media During the Arab Spring?" available from *philhoward.org/?p=789.*

34. "How Egyptians Used Twitter during the January Crisis," available from *mashable.com/2011/01/31/egypt-twitter-infographic/*, accessed June 3, 2015.

35. "Twitter and Privacy: Nearly One-in-Five Tweets Divulge User Location through Geo-Tagging or Metadata," available from *https://pressroom.usc.edu/twitter-and-privacy-nearly-one-in-five-tweets-divulge-user-location-through-geotagging-or-metadata/.*

36. S. Goel and D. Goldstein, "Predicting Individual Behavior with Social Networks," *Marketing Science*, Vol. 33, No. 1, 2014, pp. 82-93.

37. "Can Social Media Predict Behavior?" *Huffington Post*, July 16, 2013, available from *www.huffingtonpost.com/jane-adams/can-social-media-predict-_b_3601192.html*, accessed June 6, 2015.

38. Why We Can't Use Social Media to Predict Riots," available from *www.fastcompany.com/3014283/the-code-war/why-we-cant-use-social-media-to-predict-riots*, accessed June 8, 2015.

39. Mohammad-Ali Abbasiy, Sun-Ki Chaiz, Huan Liuy, and Kiran Sagooz, "Real-World Behavior Analysis through a Social Media Lens," 2012, available from *www.public.asu.edu/~huanliu/papers/SBP12Lens.pdf*

40. Nathan Kallus, "Predicting Crowd Behavior with Big Public Data," 2014.

41. Sitaram Asur and Bernardo A. Huberman, "Predicting the Future With Social Media," available from *www.hpl.hp.com/research/scl/papers/socialmedia/socialmedia.pdf*, accessed May 6, 2015.

42. Professor Kalev Leetaru, *Culturomics 2.0: Forecasting Large-Scale Human Behavior Using Global News Media Tone in Time and Space*, Urbana-Champaign, IL: University of Illinois, 2011.

43. "Social Media Sparked, Accelerated Egypt's Revolutionary," *Wired Magazine*, November 2, 2011, available from *www.wired.com/2011/02/egypts-revolutionary-fire/*.

44. "Social Media Did Not Create ISIS, Nor Did It Spark Revolutions," November 5, 2014, available from *english.alarabiya.net/en/views/news/middle-east/2014/11/05/Social-media-did-not-create-ISIS-nor-did-it-spark-revolutions-.html*, accessed June 3, 2015.

45. "Social Media Use Evolving in Egypt," July 4, 2014, available from *www.dw.de/social-media-use-evolving-in-egypt/a-16930251*, accessed May 3, 2015.

46. *Social Quorum Heuristic: Feasibility Study DARPA and SCL*, October 23, 2012.

47. "Social Media Sparked."

48. "Paidcontent" was an online media hub that covered news, information, and analysis of the business of digital media. It was founded in 2002 by journalist Rafat Ali.

49. "Social Media Sparked."

50. "Abrams M1A2," *Defense Industry Daily*, February 4, 2014, available from *www.defenseindustrydaily.com/digital-abrams-the-m1a2-sep-program-updated-02834/*.

51. *The Most Expensive War in World History: Costs of Iraq, Afghanistan Wars Could Rise to $6 Trillion*, available from *www.globalresearch.ca/the-most-expensive-war-in-world-history-costs-of-iraq-afghanistan-wars-could-rise-to-6-trillion/5329432*, accessed February 17, 2015.

52. Professor Julian Lindley-French, Senior Fellow, Institute of Statecraft and Director, Europa Analytical, Private email distribution, April 4, 2014.

53. William R. Polk, "American Grand Strategy for Iraq, Syria, and the Region," *The Atlantic*, June 23, 2014, available from *www.theatlantic.com/international/archive/2014/06/william-r-polk-on-grand-strategy-iraq-and-syria/373221/*, accessed May 3, 2015.

54. Author's private diary, December 2009.

55. "Force without Force," *The New Republic*, April 23, 1999, available from *www.newrepublic.com/article/politics/force-without-force*.

56. Author's quote.

U.S. ARMY WAR COLLEGE

Major General William E. Rapp
Commandant

*****

STRATEGIC STUDIES INSTITUTE
and
U.S. ARMY WAR COLLEGE PRESS

Director
Professor Douglas C. Lovelace, Jr.

Director of Research
Dr. Steven K. Metz

Author
Dr. Steve Tatham

Editor for Production
Dr. James G. Pierce

Publications Assistant
Ms. Rita A. Rummel

*****

Composition
Mrs. Jennifer E. Nevil

Made in the USA
Monee, IL
04 November 2026

67379789R00039